British Electric Trains in Camera

The Glasgow electric services have been much expanded. In 1979 a new station was commissioned at Partick to replace Partick Hill, a little to the west. Direct interchange is available from here to the Underground system. No 303.014, departing on a Springburn-Balloch service, meets No 314.214 on a Dumbarton Central-Motherwell train on 18 August 1980. The station buildings on both platforms consist of yellow brown transparent domes, with seats inside, and escalators down to streel level. *Cyril Lofthus*

British Electric Trains in Camera

John Glover

LONDON

IAN ALLAN LTD

Cover: Unit No 308.150 at Witham with the 14.22 from Braintree on 2 August 1981. *Les Bertram.*

Contents

First published 1982

ISBN 0 7110 1163 X

Published by Ian Allan Ltd, Shepperton, Surrey and printed in the United Kingdom by Ian Allan Printing Ltd at their works at Coombelands, Runnymede, England.

Introduction

The electric multiple-unit, or emu, is perhaps the most mundane of all railway rolling stock. Yet such vehicles account for nearly half of British Rail's passenger carrying capacity (7,588 out of a total of 17,042 vehicles at the last count), and the fleets of London Transport (4,353 cars) and the Passenger Transport Executives of Tyne & Wear (90 articulated units) and Strathclyde (33 cars) are exclusively electrically powered. In view of the importance of the emu, it is surprising that more pictorial attention has not been given to the subject. This volume depicts the British scene at the beginning of the 1980s.

In this country the electric multiple-unit has been essentially a means of urban transit. Even though steam traction was used on the Metropolitan Railway in its early days, it was far from ideal, and quite impossible for use in tube tunnels. Electricity was thus the chosen means of propulsion, and the City & South London Railway of 1890 and the Central London Railway of 1900 both used electric traction – but with locomotives. Locomotives however had their problems, one of which was the time taken in terminal manoeuvres. On the CLR this restricted the minimum service interval to 3½min, which was promptly reduced to 2min with multiple-unit working. Another difficulty was that the performance varied with the number of vehicles being hauled. With the multiple-unit system each unit has a fixed power/weight ratio, and the same ratio applies to a train formed of two or more similar units coupled together, the train therefore having the same performance characteristics as a single unit.

Multiple-unit development took place initially in the United States as an extension of tramcar technology. The American influence was clearly visible in many early vehicles, and the continuing use of the word 'car' in preference to 'coach' is an enduring reminder of transatlantic origins. There were three essential pre-requisites for successful electric traction: development of a sufficiently reliable and powerful electric motor, an adequate means of current collection by a moving train, and a suitable means of control; for multiple-unit working this meant simultaneous control of all the traction motors throughout the train from a single master controller. Achievemenof this last requirement is credited to Frank J. Sprague of Chicago in 1897, and from that time onwards urban electric railways spread rapidly.

Several British railways showed an interest, and by 1914 places as diverse as Morecambe, Birkenhead, Holcombe Brook, Tynemouth, Crystal Palace and Peckham Rye were accessible by electric train from at least one direction. By this date too, all the in-town lines of what subsequently became London Transport were also electrified. As might be expected in these early years of experiment, electrification systems varied widely. They were thus incompatible, but as there was no physical connection between them, this was unimportant. Thus ac vied with dc, third (or fourth) rail with overhead current collection, and so on. New schemes were inaugurated throughout World War I and into the Twenties, and it was during this period that the Southern Railway adopted the London & South Western Railway's third rail dc electrification system and began the expansion which was to bring large tracts of Southern England into live rail territory in the ensuing 20 years.

During the inter-war period, technical opinion moved towards standardisation and a recommendation for future construction was made. The standard was to be 1,500 volts dc with overhead power supply, a combination not then seen in Britain for passenger railways. It was adopted by the London Midland & Scottish Railway, which electrified the 8¾ miles between Manchester and Altrincham in 1931. The London & North Eastern Railway was also interested and planned to electrify both Liverpool Street-Shenfield and Manchester-Sheffield-Wath, but inauguration of electric services was delayed until British Railways days and took place in 1949 and 1954 respectively.

The principle of 25kV ac overhead electrification came to this country from France in the early 1950s, and offered considerable advantages. Overhead contact wires could be of smaller cross-section, with savings not only in the cost of copper but also in the reduced dimensions of the steelwork needed to support it. Fewer feeder stations for electric power were required, and there was a promise of greater adhesion. Trials were authorised, and in 1952 the Lancaster, Morecambe and Heysham line became the first in the United Kingdom to be electrified with industrial-frequency (50Hz) ac. The decision to adopt ac in future (except in Southern Region territory) was confirmed, notwithstanding the increased clearance between live equipment and fixed structures necessary for safety at the high voltage. Although the clearance requirements were reduced in 1962, considerable sections of line in the London area of

In order to supplement the local Coventry service, four additional class 310 units were built in 1975. Numbered 312.201-4, they resemble the earlier units; one change is that they are gangwayed throughout whereas the Class 310 has no connection between the centre pair of cars. The cab fronts are fitted with flat windows, enabling toughened glass to be used besides being cheaper. No 312.201 was photographed at Wolverhampton on 6 January 1979. *Peter Shoesmith*

the Eastern Region and in parts of Glasgow where providing the original clearance requirements would have been prohibitively costly, had already been electrified at the reduced voltage of 6.25kV. All the 6.25kV sections still remaining are expected to have been converted to the full 25kV by 1984.

The 1955 Modernisation Plan foreshadowed large-scale electrification, particularly in the 'wide and accepted field of suburban services', and 3,600 emu vehicles were to be built. Apart from substantial extensions on the Southern Region to the Kent Coast and later to Bournemouth which perpetuated the third rail, all the new work was to the new ac standard. Subsequent schemes have also been at 25kV ac, save only the minor third rail extensions on Merseyside and the specialised dual-voltage Great Northern service which utilises third rail between Drayton Park and Moorgate.

Turning now to the trains themselves, three distinct phases of evolution from the passengers' viewpoint can be seen in the multiple-units in service today. First, there are the relatively few types of stock remaining which date back to the development era of company designs. Particularly notable here are the lightweight units with sliding doors for the Southport and Wirral services of the LMS, and the 1938 stock of London Transport, which was the first tube design to pack all the power equipment underneath the passenger accommodation, and thereby maximise the space available for passengers. Less distinguished are the many heavy Southern suburban trains with a mixture of open saloon and compartment seating. All have slam doors and a dreary interior design whose only positive attribute is the seating of the maximum number of passengers in the minimum possible space.

The second phase came with the Modernisation Plan. To meet the operators' needs, mass production of BR 'Standard' designs was authorised; for better or worse these largely echoed Southern practice with their emphasis on seats for all. Just as the Mark I locomotive-hauled stock failed to impress, so too did the new generation of emus. In retrospect it is easy to be critical, but even the Plan itself admitted that there were design problems. 'Research has been undertaken into the design of bogies for electric multiple-unit vehicles', it said, 'and it is confidently expected that future designs will provide an entirely satisfactory standard of passenger comfort'. Alas, another 10 years were to elapse before that particular problem was finally solved. Two types of units from this era stood out; curiously these were also the only two which were not smothered externally in drab green paint. The celebrated Blue Trains of Clydeside, although initially a technical disaster, featured bright and open interiors with good all-round vision for passengers; and the maroon painted Clacton express stock for the Great Eastern combined the best features of the final Mark I coach designs with a stylish exterior. All the trains of this vintage are now 20 years old or so, and a refurbishing programme to equip them for another 20 years of life is now well under way.

From about 1965 a change became apparent. Although traditional designs continued to be churned out on behalf of the Southern until 1974, the appearance of the AM10 (now Class 310) units on the London Midland Region demonstrated that a 'back to fundamentals' approach was now being adopted. At last, a good riding, comfortable and modern vehicle was being produced. This was followed by the PEP train of 1971 for inner suburban use; after some years of gestation it eventually spawned five derivatives for the Great Northern (Class 313), Merseyside (Class 507), Glasgow (Class 314), Southern (Class 508) and Shenfield (Class 315). Here indeed was a revolutionary approach, with automatic couplers among the many features included as standard.

New thoughts were apparent at London Transport too, where the 1950s and early 1960s had been spent improving the prewar designs. One-man operated trains for the Victoria Line came with the 1967 stock, and a trend to longer but fewer cars was seen in both the 1973 stock for the Piccadilly and the D stock for the District Line. Both the PTE designs offered a new deal to the passenger, with the Tyne & Wear Metrocars owing much to Continental ideas on light rail technology.

What of the future? For British Rail it would seem that the day of the electric train *as such* is passing. The Advanced Passenger Train is an engineering concept which happens at present to be electrically powered; the experimental train was gas turbine driven, and subsequent production could have diesel engines fitted with relatively little structural alteration. In the traditional fields of electric units, the new Class 317 trains for the Midland Suburban Electrification are based on the Mark III coach, as are the Class 210 diesel-electric prototypes. A train is thus being designed for a job in the commercial and operating sense. Any system of propulsion can then be installed, which seems an eminently sensible approach. After all, it is of no concern to the travelling public and only of limited interest to the operator as to what makes a train move; the important point is that it should do so, and in accordance with an acceptable timetable.

Readers will find little reference to the technological element in this book, but this has already been very adequately covered by B. K. Cooper in his *Electric Trains in Britain* (Ian Allan, 1979). In compiling this album, I am fortunate to have been able to draw upon the work of a number of experienced railway photographers, whose efforts have enabled a comprehensive picture of present day operations to be portrayed. I am grateful to them all.

Ewell, Surrey
February 1981 *John Glover*

Southern Region

Left: If the Southern Region epitomises the traditional electric railway, then Clapham Junction epitomises the Southern. It is also the point where the South Western and Central Divisions (no connection!) run side by side. On the left, Wimbledon-based Class 405 (4SUB) unit No 4643 leaves for Shepperton, while sister Class 405 unit No 4749 from Selhurst forms a Beckenham Junction service in October 1977. *John Glover*

Below left: The SUB units which remain in service date from 1948, but South Bermondsey station is clearly rather older – illustrating the fact that it is not only in trains in which investment is needed. 4SUB No 4293 is the leading unit entering South Bermondsey with a London Bridge-Wimbledon-Sutton-London Bridge circular service on the dull morning of 7 May 1980. *John Glover*

Above: No 4124 is one of an earlier batch of Class 405s which have now all been withdrawn. The unit is seen in Clapham Cutting on 15 August 1973, forming the 11.06 Waterloo-Chessington South. The prominent metal route stencil used to be one of the hallmarks of Southern multiple-units; now roller blinds have taken over on all other stock, and even a few of the 4SUBs are now fitted with them. *Brian Morrison*

Left: 4SUB units can only work in multiple with each other as they have different control gear from more recent vehicles. They are also restricted to the parts of the system electrified at 660V. This picture of two recently outshopped units, Nos 4283 and 4677, was taken at Coulsdon North, whence they were to form the 09.40 to Victoria on 4 November 1980. *John Glover*

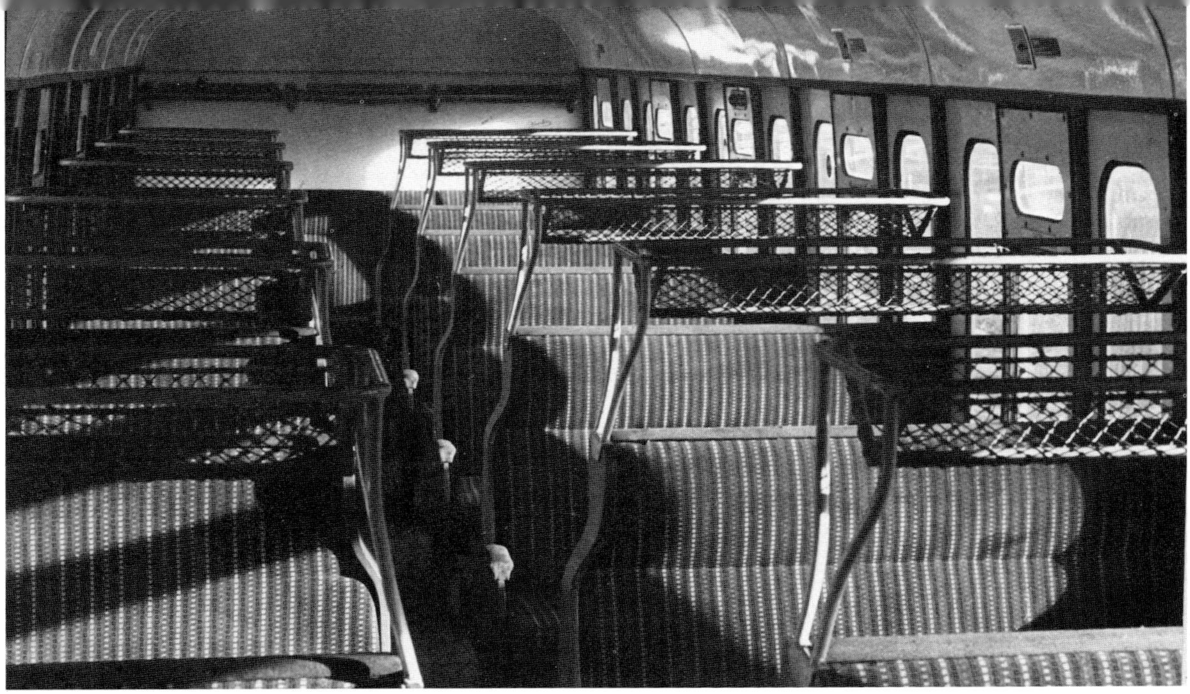

Top: This interior view is of a saloon in a 4SUB Motor Brake Second, but the 3+2 seating either side of a central gangway with luggage racks above the seats is typical of all Southern suburban units, of which the design changed little during their entire period of construction. *John Glover*

Above: The SUB units are identifiable by their front end design, but the trains that followed all have an uncanny similarity with each other. No 5143 is one of the 4EPB units of Class 415/1 introduced from 1951 with bodywork to Southern Railway design; the principal distinguishing feature is the quarterlight in each door, which is missing on BR-designed stock. No 5143 had been newly painted in blue and grey and is seen at Selhurst Depot Open Day on 21 September 1980. The WR diesel unit on the left is, of course, a visitor. *John Glover*

Left: No 5194 is another unit of Class 415/1 and is seen here forming the 11.33 Dartford-Cannon Street via Sidcup, and arriving at New Eltham on 11 September 1978. The Southern has always made bicycle storage facilities available to passengers, as can be seen by the well patronised shed on the right. *Brian Morrison*

Below: Sunnymeads, on the Windsor branch, is a delightful unstaffed station consisting of a single island platform reached by this footbridge. 4EPB No 5107 is arriving with an afternoon Waterloo-Windsor & Eton Riverside service on 23 August 1977. The arrangement of the three panels forming the front of the unit is a recognition feature for Southern Railway as opposed to British Railways designs. *John Glover*

5206
55

Far left: With buckeye coupling at the ready, EPB units No 5206 and 5192 form the 09.51 Cannon Street to Gravesend near Mottingham on 22 August 1979. There is a remarkable amount of clutter on the roof, bearing in mind that none of it relates to water supply for toilets. *Brian Morrison*

Above left: A characteristic of the British Railways designed Southern stock is the slight roof overhang on the cab, as can be seen in this wintry view of Class 415/2 (4EPB) No 5309 leaving Chislehurst Tunnel on 29 December 1976 with the 13.07 Cannon Street-Orpington. These units date from 1960, and this is one of the few of the class fitted with express gear ratios, permitting them to run at 90mph. All are allocated to the South Eastern Division. *Brian Morrison*

Centre left: Class 415/2 No 5365 is undergoing overhaul at Chart Leacon, Ashford, on 5 September 1973. This is one of the later units of the batch with smaller headcode blind panels. The general similarity of these trains to their diesel-electric counterparts is illustrated by Hampshire unit No 1110 of Class 205 on the left. *Brian Morrison*

Below: Turning to the two-car units, Class 418/0 are 2SAP units which were built as 2HAPs with first class accommodation in 1957 but subsequently downclassed. The Southern Railway design bodies were constructed on underframes from withdrawn 2NOL units of 20 years previously. The driving trailer has a side corridor giving access to a single lavatory from the eight compartments. This is unit No 5615 at Wimbledon on 25 June 1980, with a service for London Bridge. *Ian Cowley*

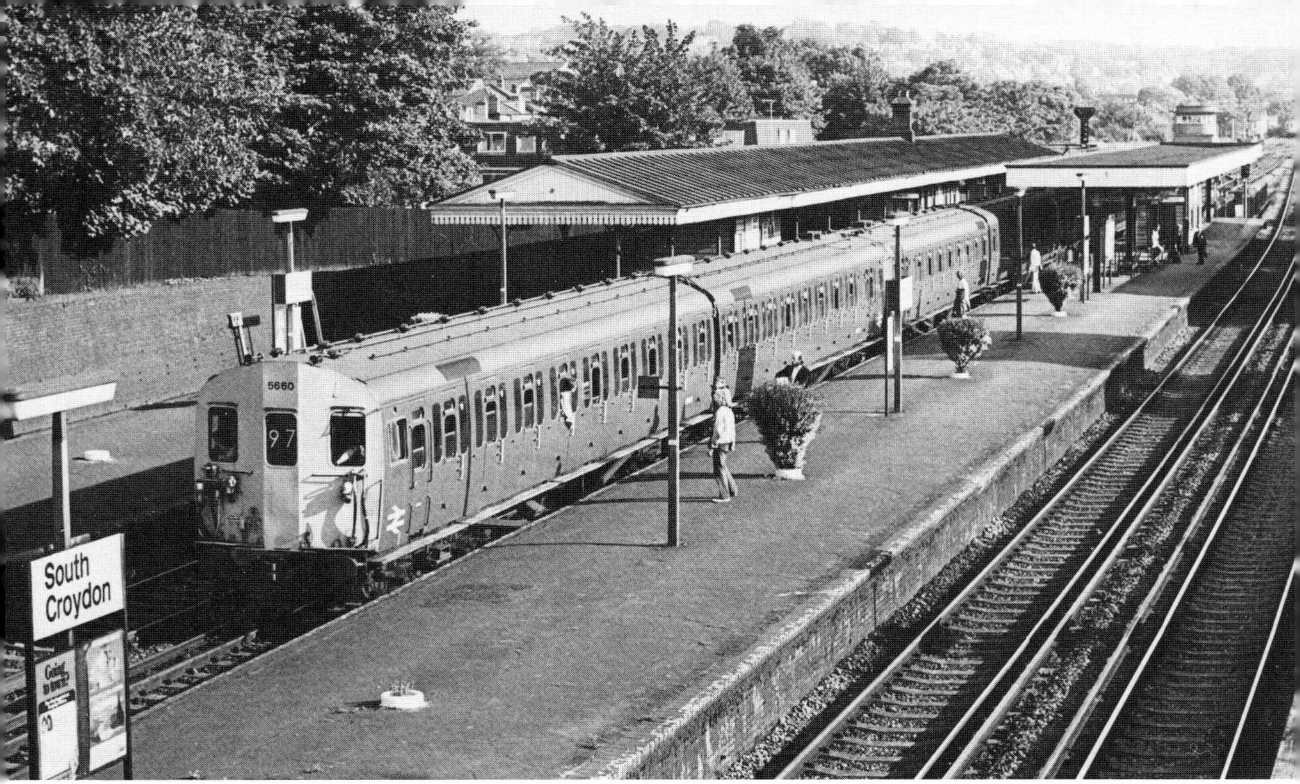

Above: Also of Southern Railway design are the 2EPB units of Class 416/1. Built in 1959, these too were new bodies on old frames. Unit No 5660 with Class 415/1 No 5017 behind arrives at South Croydon with the 17.50 Tattenham Corner/18.01 Caterham-London Bridge on 3 July 1980. These two trains combined at Purley. *Les Bertram*

Below: Facilities at the Tattenham Corner terminus are more than adequate for the traffic offering, except perhaps on Derby Day. Class 416/1 No 5666 arrives with the 14.50 from Charing Cross on 25 March 1980. The inverted black triangle is an indication to station staff that it is this vehicle which contains the guard and luggage, so that it can be located quickly should there be any parcels traffic to be dealt with. *Brian Morrison*

Above: Morden Road is unstaffed, in spite of appearances. It is situated on the Wimbledon-West Croydon line, and unit No 5665 is approaching on the bright but cold morning of 1 March 1979 with a train for West Croydon. This line is single track, with passing loops at Mitcham Junction and Waddon Marsh. A half-hourly service is provided six days a week. *John Glover*

Left: British Railways were responsible for the 2EPB rolling stock in the 57xx number series. This was built from 1954, and unit No 5721, now Class 416/2, leaves Coombe Road for Elmers End on 31 October 1980 on the threatened peak-hour only service from Sanderstead. It is about to enter Coombe Lane, Park Hill and Woodside tunnels, the two short daylight gaps between them being hardly visible. The wide centre panel on the vehicle ends of the BR-designed stock is apparent in this picture. *John Glover*

Left: On a cold, wet and generally unpleasant morning in February 1978, 2EPB No 5738 enters Woodside station with the shuttle train to Addiscombe. It is about midday, but the clock has long since stopped; on the left can be seen the white paint on the station building which was previously used as a background for the semaphore signals to make sighting easier. *John Glover*

Below: The last 15 units of Class 416/2 were constructed at Eastleigh for the Eastern Region's Newcastle-South Shields service. After de-electrification in 1963, these units were reallocated to the Southern Region. They are distinguishable by the enlarged brake compartment, which means the loss of one seating bay of 10 seats. Presumably Tyneside folk were more inclined to bring their prams and bicycles than the average traveller on the Southern! An unidentified unit passes Tolworth Tower with the 09.36 Waterloo-Chessington South on 19 January 1980. *Colin Marsden*

Above: The other 'different' feature of the ex-Newcastle trains is the smaller headcode panel. No 5794 is the rear unit leaving Tolworth on the 13.35 Chessington South-Waterloo on 25 March 1980. The steep gradients of this line, built for electric traction in 1938, are clearly visible. *Brian Morrison*

Left: The final non-gangwayed units of 2HAP are really express units. Introduced in 1957, they are now to be found on all three Divisions. Class 414/3 units Nos 6135 and 6152 arrive at Sandling on 12 June 1976, forming the 12.30 Charing Cross-Margate via Deal. The driving trailers are composite vehicles with separate lavatories in the centre of the coach for first and second class passengers respectively. *Wyn Hobson*

Left: At Longfield on 4 October 1980, 2HAP unit No 6094 leads a Victoria-bound train which is attracting substantial off-peak patronage. The cleaner roof line of these more modern units may be compared with the earlier Southern Railway designs. *John Glover*

Below: This is Sunnymeads again, immediately on the London side of the station on 8 May 1980. Few of the weekend yachtsmen, one imagines, travel by train. The 15.41 Waterloo-Windsor & Eton Riverside is approaching, with 2HAP unit No 6043 leading. This was one of the units downgraded to all-second class in 1977 as a 2SAP, only to be converted back again in 1980. *Les Bertram*

Left: In second class only guise, 2SAP unit No 5943 leaves Farnborough Main at sunset at the head of the 17.54 Waterloo-Eastleigh on 11 April 1980. The footboards along the sides of the unit span several doorways, although this picture demonstrates the limited areas upon which passengers actually step. *Brian Morrison*

Below: Class 414/3 2HAP No 6136 stands in Selhurst Depot with rather undignified lines of bunting attached to its front handrails during the Open Day on 21 August 1980. It has just been repainted in the now standard blue and grey livery; in due course all other suburban units are intended to follow suit, except for the 4SUBs, which have only a short life expectancy. *John Glover*

18

Far left: Specially designed express rolling stock for the Kent Coast services was delivered from 1958 onwards, following prototypes completed in 1956. Known as 4CEP (or 4BEP for the buffet version) they closely resemble Mark I coaching stock, with extensive use of Formica panelling. Class 411/2 No 7178, forming the 13.23 Ramsgate-Charing Cross via Deal, passes Buckland Junction, Dover, on 31 March 1975. *Wyn Hobson*

Above: Sandling is a favourite spot for photographers. On the left is the gated and disused branch to Hythe; it extends only a few hundred yards. Class 410/2 4BEP unit No 7015 is arriving at the head of a Charing Cross-Margate train via Deal on 24 August 1979. *John Glover*

Left: Buffet car sets are not often seen on their own, as their commercial justification requires a trainload of passengers as potential customers. However, Royal Train duty is a different matter, and a polished 4BEP, No 7019, forms a Victoria-Chichester 'Royal' on 10 March 1978. It is seen to the south of Mitcham Junction. *John Scrace*

Right: The towers at Cannon Street are all that is left of the overall roof which was finally demolished in 1956. 4CEP No 7207 was one of the last of the batch and destined for the South Western Division, but it was soon transferred to the South Eastern. All the later builds were mounted on the heavy but effective Commonwealth bogies. The train will shortly be forming the 17.56 to Margate on 7 June 1976. *Brian Morrison*

Below: The Kent Coast electrification benefitted the Continental Boat Trains. No S68001 was the first of 10 Class 419 Motor Luggage Vans built for these services. By use of their traction batteries, they are able to work on the quaysides away from the third rail. Headcode 74 denotes a service to Victoria via Meopham, but 4CEP No 7124 will also take the same route; it will form the ordinary hourly stopping train from Dover Western Docks, seen here on 24 August 1979. *John Glover*

Top left: Refurbishing of over 5,000 emu vehicles is planned to be completed by the end of this decade. After experimental work carried out in 1975, No 411.506 became the first production unit to be outshopped by Swindon Works. It is seen here at Waterloo for Press inspection on 18 December 1979. Only the bodyshells, external doors, underframes and motor bogies are from the original unit; all but one second class compartment has been eliminated, and two guard's brakes have been replaced by one in the centre of the train. *Colin Marsden*

Centre left: New seats similar to those in Mark III coaches have been installed, and this view of a Motor Second Saloon also shows the large hopper type opening windows. On the debit side, all curtains have been removed in the second class – a previous feature of Southern main line stock. Not all refurbishing work will be as extensive; EPB cars are only having compartments converted to open saloons, with new ceilings and strip lighting installed. *John Glover*

Below: The Class 421 (CIG) express stock was a 1964 development, and originally intended for the Brighton line. Class 421/1 No 7304 is the leading set of a Victoria-Horsham-Bognor Regis train on 4 September 1979. It is seen here passing Merstham on the original main line. The Redhill avoiding (Quarry) line passes under the bridge behind the signalbox. *John Glover*

Right: Although nominally express stock, the 4CIG units are often found on local services. Class 421/2 No 7362 approaches Falmer Tunnel with the 12.44 Brighton-Eastbourne on 2 April 1978. This is one of the second batch built for the Portsmouth line services in 1970, but differences are negligible. *Les Bertram*

Below: Salfords station has platforms on the slow lines only. Class 421/2 units Nos 7369 and 7363 are passing with the 14.49 Portsmouth Harbour-Victoria on 10 May 1980. There is a Driving Trailer Composite at each end of these units, with the first class accommodation in either three or four compartments. The vehicle nearest the camera is one of the three-firsts variety, the fourth compartment being second class. Changes could be made in future should more (or less) than the 42 first class seats provided in a unit be required. *Les Bertram*

Left: Chichester station was reconstructed during the Modernisation Plan era and completed in 1961. Similarities to many of the stations rebuilt in Kent at the same time are apparent, and it is already a period piece in its own right. 4CIG No 7335 pauses with a fast service to Victoria via the Mid-Sussex line on 9 August 1980. *Les Nixon*

Below: A 12-car CIG/BIG/VEP formation is seen at Brighton on 11 August 1979. In the extreme foreground are the platforms for the Worthing line services, while a train from Eastbourne has crossed the viaduct in the background. The distinctive signalbox will be swept away under the resignalling scheme. In the foreground are locomotives Nos 33.023/52/21, together with Nos 73.101/4. A quite remarkable quantity of huts of varying shapes and sizes is also visible. *Gavin Morrison*

Passengers mus
not cross the lin
Notice
Persons in charg
of vehicles must
obtain signalma
permission by
telephone befor
using the crossi

Top left: A light dusting of snow on 11 April 1978 has transformed Clapham Cutting, a favourite vantage point. A rather grimy Class 423 (4VEP), No 7778, heads a Victoria-Bognor Regis service. External washing of trains has to be suspended in frosty weather due to washing installations becoming frozen, but also to guard against patches of ice forming on footboards. *John Glover*

Left: This is Brighton as seen from the Worthing line platforms on 20 July 1978. 4VEP No 7759 with a 2HAP unit behind is arriving with a Coastway train from Littlehampton. These 4VEP units have a similar general layout to the 4CIG units, but because of a higher density configuration can carry 20% more seated passengers. *Les Nixon*

Above: Early morning mist at Surbiton on 1 June 1979 sees a 4VEP unit depart for Waterloo on the up slow line with the 06.21 train from Guildford. The telephoto lens has accentuated any slight imperfections in the trackwork, which are rarely noticeable in real life! *John Glover*

Above: Another preoccupation of photographers is with finding good – and preferably new – vantage points. This splendid overall view of Eastbourne station shows the irregular shape of the platforms, one of which is sufficiently curved to need 'Mind the Gap' warnings to be painted on the platform edge. Class 423 No 7853 is leaving on 28 August 1977 with the 11.53 to Brighton. This picture also shows the large area of land devoted to railway purposes in the town.
Les Bertram

Right: The 4VEP units are found all over the Southern Region, and among their duties are weekend work on the Brockenhurst-Lymington Pier branch. No 7826 is leaving Brockenhurst in September 1977. Two white blanks are shown in the headcode panel for this local movement, as a headcode is deemed unnecessary. *John Glover*

Right: In 1978 12 4VEP units were specially adapted for services between Victoria and Gatwick Airport. End seats in the second class saloons were removed and replaced with luggage racks, and the units were specially marked. They are known as 4VEG stock, or Class 427. The windows next to the racks are fitted with bars and have been painted over externally, as can be seen in this view of No 7906 leaving Gatwick Airport for Victoria with the 16.40 from Brighton on 23 March 1980. *Les Bertram*

Below: A 12-car formation makes a substantial train about 258yd in length. A Class 430 (4REP) power unit draws two unpowered Class 491 4TC units through the New Forest near Beaulieu Road on a Weymouth-Waterloo working in May 1977. Four 365hp traction motors are installed in each of the end vehicles of the REP unit, resulting in a nominal 2,920hp for the entire train; this compares with 3,000hp for a conventional CIG/BIG/CIG formation. *John Glover*

Bottom: A semi-fast Bournemouth-Waterloo service arrives at Brockenhurst on 21 July 1979. 4REP No 3003 is the motive power. All REP units include a buffet car (the R in REP stands for Restaurant), and this is the third vehicle from the front in the picture. It was converted from a locomotive-hauled vehicle and in this unit is now named *The Farnborough.* *Gavin Morrison*

Above: A short, sharp thunderstorm at Winchester catches the 13.46 Waterloo-Bournemouth in the summer of 1976. The Bournemouth trains were originally painted in all-over Rail Blue livery; this example is blue and grey, although the grey is barely visible. The TC trailer sets were converted from locomotive-hauled stock, and even in the REP powered units only the two motorcars were of new construction. *John Glover*

Below: At Bournemouth a Class 33/1 diesel backs down on to the front of through trains, and takes either one or both of the 4TC trailer units on to Weymouth. It then propels these back to Bournemouth and on to a waiting 4REP tractor unit. Here an eight-car formation is approaching Poole on 29 August 1978 with the diesel locomotive in the rear. It is passing No 47.487, waiting to form an inter-Regional train via Reading, and No 73.125. *Gavin Morrison*

Above right: When Royalty travels, all sorts of curious things can happen. Electro-diesel No 73.142, now named *Broadlands,* is conveying HM Queen Elizabeth the Queen Mother from Portsmouth to Waterloo in a train consisting of the driving trailers from 4TC unit No 416, with the brake and buffet trailers from 4REP No 3009. This buffet car has the stirring title of *The Wimbledon.* The train is approaching Clapham Junction on 2 August 1979. *Brian Morrison*

Left: The appearance in 1971 of the prototype PEP units marked a new departure for inner suburban services. Originally two four-car and one two-car trains were constructed, and were placed in passenger service on the Southern. Unit No 4002 subsequently became a bogie and motor test unit; No 920.001 was formed of two of the original Motor Open Brake Seconds, together with a new Trailer Second, and converted to ac traction. Both units are seen here at Derby Technical Centre on 6 April 1979. *Colin Marsden*

Bottom left: No 4002 returned to the Southern and was photographed on test approaching Surbiton on the down slow on 18 June 1980. It is finished in a striking livery, akin to the APT-P, of silver-grey and black with a red stripe. These prototype vehicles are very different from the production variety. Styling differences include the flat front of all the derivatives; the Scharfenberg couplers and the original type of traction motors were among the features discarded. *John Glover*

Left: Internally there was a considerable revision of ideas in the PEP stock. Seating was arranged in a 2+2 formation as seen here, and there were three pairs of double doors on each side of the intermediate vehicles (two on the driving ends). Seating capacity for the four-car unit at 268 was thus appreciably less than the production Class 508 units, which have 316 seats. *John Glover*

Below: This Class 508 interior offers an interesting comparison. The seat design itself is entirely different, and the handgrips have been replaced by additional handrails. On the Class 508s (and also on the Class 315s), passenger control of doors is arranged by pushbutton, and passengers may close doors themselves from the inside only. To reassure the nervous, it is *not* possible to prevent a door from being opened from the outside by keeping one's hand on the 'Close' button inside! *John Glover*

Right: The Class 508s are the first stock on the Southern not to display a headcode. Instead, two blinds are fitted, one above each cab window. No 508.013 is leaving Waterloo on 13 May 1980 with one of the three peak hour-only trains to Shepperton via Richmond; the normal route is via Wimbledon. A variety of multiple-units can be seen in the station, together with a Class 73 and a Class 50. *John Glover*

508 021

Effingham Jn
Waterloo
508 043
Ewell West

Above left: On 14 April 1980, Nos 508.021/2 are seen at New Malden with the 14.34 Waterloo-Shepperton. In the up direction No 508.009 is on a driver training trip. These trains are fitted with the Tightlok coupler, as are all the production versions of this design. *Colin Marsden*

Left: Forty-three trains of Class 508 have been produced. They were originally destined for Merseyside as three-car units but were diverted to the Southern and had an additional trailer inserted. The final Class 508 unit, No 508.043 is at the front of an Effingham Junction-Waterloo train as it arrives at Ewell West on 22 October 1980. *John Glover*

Above: At Epsom cross-platform interchange is available between trains bound for Dorking and Effingham Junction. On 22 October 1980 No 508.039 comprises a Waterloo-Effingham Junction service; on the right is 2SAP unit No 5609 of Class 418/0 forming a Victoria-Dorking-Horsham service. Above the tracks is Epsom's celebrated signalbox. Forty-three new trains cannot on their own modernise a railway system. *John Glover*

Centre right: Whilst the majority of old vehicles are condemned, a few are retained for departmental use. The two cars of de-icing unit No 001, photographed in Bournemouth depot sidings on 27 March 1977, were originally driving motor coaches of a 2HAL and a 4LAV. Although only one bogie in each car has traction motors, shoebeams and pick-up shoes are fitted to all bogies to minimise the risk of stalling in icy conditions. *Brian Morrison*

Bottom right: Former 4SUB Driving Motor Coaches make up Stores Unit No 024, which is approaching Farnborough Main on the up fast on 11 April 1980. Together with two other similar units, it is used to convey stores between works and depots, and is probably en route from BREL Eastleigh. *Brian Morrison*

Above: The Waterloo & City line rolling stock consists of 12 Driving Motor Coaches and 16 trailers. Normally, trains have a pair of motor coaches, and three trailers sandwiched between them. Class 487 motor car No S62 with trailer S86 behind awaits departure from Bank on 8 December 1979. *Brian Morrison*

Centre right: Internally the cars are rather cramped, and their design does not show the same flair as the contemporary stock of London Transport. This official view was taken in 1939 when the cars were new, but little has changed since. The ventilator grilles in the far end bulkhead read 'Southern Railway', and some of these are still in existence. *British Rail*

Bottom right: At Waterloo, arriving trains run forward into the depot sidings before setting back into the other platform for the return working. The entire sidings complex is underground, and in normal service the trains never see the light of day. This view was taken on 8 December 1979. *Brian Morrison*

Top: To reach the surface at Waterloo, the 49ft cars have to be raised in a hoist. Since they have the same electrical system as the rest of the Southern stock they may run under their own power on surface tracks, but this is a rare sight. Motor car No S59 was photographed on 21 August 1980 undergoing overhaul in Selhurst depot; its diminutive size can be judged from the vehicle behind it. *John Glover*

Above: In order to provide train services on the Isle of Wight with stock which could meet the severe height restrictions, former London Transport pre-1938 tube stock was purchased. After overhaul by its former owners, conversion to third rail and the fitting of luggage racks in place of some seats was carried out at Stewarts Lane. Prior to delivery to the Island, a VEC+TIS formation was photographed passing Weybridge on a crew training run on 13 July 1966. *Geoffrey Kichenside*

Above: The train service normally works in two parts. A shuttle service operates between Ryde Pier Head and Ryde Esplanade, and Class 485 4VEC unit No 042 is seen on 9 August 1980 at one end of this train in Esplanade station. The other track is used for the Pier Head-Shanklin service, which resumes normal double track operation at the far end of the station. In the foreground are the remains of the pier tramway. *John Glover*

Right: 4VEC No 046, with a 3TIS (Class 486) unit behind, sways along through the lush Isle of Wight countryside between Brading and Smallbrook on 9 August 1980. Bembridge Down can be seen in the background. In practice the location is less than idyllic; a constant stream of traffic was passing only a few inches behind the photographer! *John Glover*

Top left: Brading station sees unit No 041 at the front of an arriving northbound train on the same day. It will have to wait for the southbound service before entering the single line which extends from here to Smallbrook. The use of clerestory-roofed stock 60 years old is unique, yet since 1967 there has been only one total train failure in service. The stock is not ideal for a line which extends along a pier; each time the doors are opened, half the roof disappears! *John Glover*

Centre left: A 3TIS unit (Class 486) displays its grey painted doors – even these trains have adopted a blue and grey colour scheme – while heading for Shanklin at the midway point between Brading and Sandown. On Summer Saturdays when up to 30,000 passengers may be carried, the timetable demands a 15-minute frequency service over the 8½ miles right through to 17.00. Intensive diagramming can mean that a train will cover as much as 250 miles in a day. With three sections of single track, including both termini, precision in running is essential. This view was taken on 9 August 1980. *John Glover*

Below: The telephoto lens has produced a result strongly reminiscent of a caterpillar as 4VEC No 041 leaves Sandown for Ryde Pier Head on the same day. The fleet of 40 vehicles has now run over 3½ million miles on the Island. Maintenance is carried out at Ryde St John's Works, where the housing of the electrical equipment inside rather than beneath the body is found to make life much easier in the difficult Island conditions. Traction motors however are still sent to Acton Works for overhaul. *John Glover*

London Midland Region

Left: The dc electric services in the London area of the London Midland Region are all operated by three-car electric units (Class 501) built at Eastleigh in 1957-8. Running under clear signals, the 12.25 Richmond-Broad Street with blinds already changed for the return trip passes the old Shoreditch station, 40 years after closure. This view was taken from the former Shoreditch Town Hall on 10 July 1980. *Martin Higginson*

Below: A peak hour-only Watford Junction-Broad Street service approaches Caledonian Road & Barnsbury on 21 July 1980. Many North London line stations have been rebuilt in a similar simple style, so that ticket issue and collection can be performed by one man. These services are the only ones on British Rail where the old 'alpha' train descriptions are still used. 'B' in the headcode denotes a stopping passenger train, equivalent to today's Class 2. The number relates to the route. *John Glover*

Left: The long-term future of the decrepit Broad Street is still undecided. A Richmond train waits to depart from its sorry remains on 6 May 1978. Although the Class 501 stock has obvious affinities with Southern units, it is in contrast built on the short 57ft underframes, and the power equipment was supplied by GEC instead of English Electric, which supplied the contemporary Southern equipments. *Les Nixon*

Below: A Watford Junction-Euston train stands at the subterranean Wembley Central station in the late evening of 27 March 1980. The fourth rail is provided for the benefit of the four daily Bakerloo Line trains to and from Elephant & Castle; use of the fourth rail was discontinued by the British Rail trains in 1970. *Peter Shoesmith*

Left: The single-tracked Croxley Green branch has only a peak hour service to and from Watford Junction. A wet morning in April 1977 sees both passengers and train depart from the somewhat basic Watford West station. The redundant fourth rail is still in position. *John Glover*

Below: Merseyside also has an extensive network of third-rail services, electrified at 630V dc. In 1938 the London Midland & Scottish Railway ordered some new cars for the Wirral extensions to the electrified network, and further cars to replace the original Mersey Railway stock were built to the same general design in 1956. Both types are now known as Class 503. On 3 July 1980, a train of 1938 cars with Motor Brake Second No M28686M at the head stands in New Brighton platform, whilst on the right 1956 stock Driving Trailer Second No M29146M is in the siding. *David Flitcroft*

Right: As originally built, both sets of cars were without end doors. Conversion took place between 1972 and 1976 to satisfy Department of Transport safety requirements for operation in the single track 'Loop' tunnel to be constructed in Liverpool. This view shows one of the 1956 cars, in green livery before conversion, entering Birkenhead Park station in June 1969. Remodelling of the station since has resulted in the removal of the centre tracks. *John Glover*

CAVENDISH
LIVERPOOL CENTRAL
M28390M
WARNING
KEEP OFF
ELECTRIC
LIVE RAILS

Right: A general interior view of the 1956 vehicles as built shows a commendably clean and uncluttered appearance. The sliding doors may be opened by passengers using 'Push to Open' buttons; on the earlier stock the buttons were mounted on the doors themselves, a practice now revived on the Tyne & Wear Metrocars. *British Rail*

Below: Lime Street is one of the stations served by the Loop services introduced in 1977, and car No M29131M will shortly be leading a New Brighton-bound service on 30 June 1980. These cars do not have conventional buffing gear, and a centrally mounted buckeye coupler is deemed adequate. *John Glover*

Left: A 1977 extension to the electrified network was from Walton Junction on the Ormskirk line to Kirkby. The last 1½ miles from Fazakerley box to Kirkby are single track. A six-car train of Class 503 stock has just arrived from Garston on the same day with No M28584M in the lead. Passengers requiring onwards transport have the choice of a dmu connection towards Wigan or a bus from the interchange at street level. *John Glover*

Below: From Kirkby, trains run through Liverpool to Garston, which was reopened with electric traction in 1978. As can be seen by the disused tracks still in situ, hope of eventual extension of the service has not been abandoned. A 1956-built train will shortly be forming a Kirkby service on 3 July 1980, with No M28381M, a Motor Brake Second, leading. At this stage Kirkby had yet to be incorporated on the destination blinds, and instead a board is displayed in the cab window. *David Flitcroft*

Right: For the Southport services the LMS itself built some not dissimilar units in 1939-41, but these had ordinary British buffing gear. The Class 502 stock was formed into three- and five-car sets, but stock shortages in their final years necessitated ad hoc reformations. Motor Brake Second No M28332M has acquired an all blue centre trailer, and was photographed in Southport sidings on 28 July 1980. It is difficult to believe that it predates the Class 507 stock in the background by as much as 40 years. *Martin Higginson*

Below: A Liverpool Exchange-Southport train runs into Birkdale on 9 September 1976, with Driving Trailer No M29881M in the lead. The station surroundings are somewhat dismal, but the flower tubs suggest that someone has tried! *Kevin Lane*

Below right: St Michaels is one of the reopened stations on the former Cheshire Lines Committee line out to Garston. Besides electrification, considerable refurbishing of station buildings was necessary. No M28351M is the leading vehicle on a Kirkby-Garston train on 30 June 1980; the raised sections of the roof are air intake ducts to provide ventilation for the traction motors. *John Glover*

Above: All good things come to an end, and the passenger-carrying career of the Class 502s finished at the close of 1980. Two vehicles have been preserved, but most have gone for scrap. Driving Trailer No M29881M contemplates its fate in an East Anglian scrapyard on 3 June 1980. In due course it will be cut up and its remains deposited in mineral wagons as scrap. *Ian Cowley*

Below: Before cutting up, however, the vehicles are fired to remove all their non-metal content. This is how Motor Cars Nos M28352M and M28355M looked following this process and before the cutting torch got to work. This June 1980 picture was no doubt the last to be taken of these particular vehicles. *Ian Cowley*

Above: Replacements for the Class 502s came in the form of the PEP-derived Class 507, and an Ormskirk-bound train consisting of No 507.008 deposits morning shoppers in a downpour at Orrell Park on 30 June 1980. The first 30 of these trains were delivered from BREL York in 1978, and a further three units arrived in 1980. *Martin Higginson*

Below: No 507.004 approaches Southport from Liverpool on 23 June 1979. The signalmen in this area work fast; all signals behind the train are already restored to danger, and the two level crossings have had their gates replaced across the railway. *Les Nixon*

Left: No 507.023 pauses at Bank Hall station with an evening Southport-bound train on 30 June 1980. Cleaning of the station building has much improved the structure, which was formerly blackened with the grime of ages. Although both the Class 507s and their Class 502 predecessors are formed throughout with open saloons, the older vehicles could seat 268 as opposed to 232 in the 507 class. *John Glover*

Below: A pair of Class 507s, Nos 507.003 and 507.004 stand in Southport station on 23 June 1979. The Motor Open Brake Seconds at each end of these units are both equipped with four 110hp GEC traction motors, as are the Southern's Class 508s. The latter however have to tow an additional 32-ton trailer around with them. *Les Nixon*

Left: The Lancashire & Yorkshire Railway was determined to achieve a high-voltage dc third rail system, and to do this they encased the side-contact third rail for safety reasons. The Manchester-Bury service is thus unique in being electrified at 1,200V dc, and its stock cannot be used anywhere else. Motor Open Brake Second No M65450 stands in Manchester Victoria awaiting a punctual departure on 29 August 1980; this line records a 99% right time performance, one of the best on British Rail. *Martin Higginson*

Below: Internally, these 1959-built vehicles of Class 504 resemble strongly the Class 304 ac electric units then in course of construction for the London Midland Region. This view of a second class saloon interior was taken when the vehicles were new. All accommodation is now in open saloons, the former compartments, including the first class one, having had their partitions removed. *British Rail*

Left: The station formerly used in Bury by the electric services has now been abandoned. Car No M65444 is in the lead as the two-car unit passes Bury South signalbox at the entrance to Bolton Street station with the 13.10 from Manchester Victoria on 4 April 1978. Track simplification work has resulted in the train momentarily entering a short portion of single line before the tracks diverge again into the two platforms. *Wyn Hobson*

Below: The new £4 million bus/rail interchange at Bury was opened in the summer of 1980 by HRH Princess Alexandra. In this picture the 12.00 from Manchester is seen arriving on a snowy 17 March 1980, the actual first day of operation. The new station is much better situated than the original and by 1981 had already resulted in a 12% increase in the number of rail passengers using the line. *David Flitcroft*

Bottom: Immediately outside the new station, the electric lines cross the freight link from Castleton Junction to Rawtenstall on the level. Virtually the whole area in this photograph was once used for railway purposes; on the right is all that remains of one of Knowsley Street's station platforms. A reliveried Class 504 unit leaves for Manchester Victoria on 25 November 1980. *John Glover*

Above: Although the Manchester South Junction & Altrincham services were taken into the ac electrified network in 1971, a photographic reminder is included of these emu trains which were the first to operate on 1,500V dc overhead in Britain. Motor Brake Second No M28589M leads two three-car units into Altrincham & Bowdon terminus on 11 June 1969, when all were in green livery. *John Glover*

Above left: Still operating at 1,500V dc is the Manchester-Glossop-Hadfield service, which was electrified in 1954. Although the Woodhead route passenger services were withdrawn in 1970, the emus built in 1950 to LNER design still maintain the local service. No M59407M has its pantograph at full stretch as it stands in Glossop terminus prior to reversing for the final leg of the trip to Hadfield on 24 September 1979. *Les Nixon*

Above right: The Glossop branch is now single track, as are the spurs connecting the branch to the main line at Dinting. The 11.13 Hadfield-Manchester Piccadilly train runs on to the Glossop branch on 3 September 1980. After a 4min stand for the driver to change ends, it will return and take the line towards Dinting station. Motor Brake Second No M59404M is leading the formation. *Les Bertram*

Below: A handful of down trains in the morning peak and up services in the evening peak use the main line platforms at Dinting, omitting the trip to Glossop on the contra-peak journey and thereby saving 8min. Here the 07.15 Manchester Piccadilly to Hadfield direct train leaves Dinting on 4 August 1980. *Martin Higginson*

Above: A Manchester-Hadfield train of Class 506 stock picks its way across Dinting viaduct, upon which it appears to be somewhat precariously balanced! The design of these units closely follows the Class 306 stock built for the Liverpool Street-Shenfield electrification, and like them is formed of open saloons with sliding doors. Four 185hp GEC traction motors are fitted. The date is 23 April 1977. *Les Nixon*

Right: Another substantial engineering work is Etherow viaduct, which carries the line 136ft above the River Etherow near Broadbottom. It was built in 1842 as a timber bridge, and was rebuilt in 1859/60, with intermediate supporting pillars added later. Newly repainted No M59404M is crossing the viaduct on 4 August 1980 with a Manchester-bound train. *Martin Higginson*

Top left: At Godley, the CLC lines on the left lead to Warrington, but the suburban services continue on the right to Manchester. A Hadfield train pauses in Godley station on 22 October 1979 with Driving Trailer No M59602M leading. The heavy construction of the dc overhead equipment compared with a modern ac installation is noticeable. *John Glover*

Centre left: The guard closes the van door after a fellow railwayman has removed his bicycle from a Hadfield-Manchester train at Guide Bridge on 4 August 1980. In 1973 a cost benefit study was made of this railway to evaluate options for the future. The study found in favour of retention, with new and improved feeder bus services, new stations and higher rail service frequencies. Some progress has since been made towards all three. *Martin Higginson*

Below: Electrically-hauled freight had, however, declined to the point of extinction, to which the remains of a pair of Class 76 locomotives in Reddish depot bear sad witness. Two Driving Motor ends of the Class 506 units face the camera in this view taken on 25 February 1979. *Gavin Morrison*

Above: Both the Midland and the London Brighton & South Coast Railways pioneered single-phase ac overhead electrification, but whereas the latter's was dismantled in 1929, the Midland and its successors soldiered on with the Lancaster, Morecambe and Heysham service until 1951. The electric trains were then withdrawn, and the overhead was converted experimentally to 6.6kV, 50Hz. Three (later four) redundant dc emus from the Willesden-Earls Court service were re-equipped. Trials began in November 1952, and public electric services resumed on 17 August 1953. One of the original units with Driving Motor Car No M28221M at the front enters Lancaster Castle on 9 October 1965. *Gavin Morrison*

Above right: On the same day, a Heysham-bound train comprised of the final unit of Class 301 with Driving Motor No M28222M leading is seen near Scale Hall. The original overhead equipment here has been replaced with what became the prototypes for all subsequent ac electrification schemes. The isolation of this line, which was the reason why the Midland Railway chose it for experiment, made it similarly suitable in the 1950s. Closure came in 1966 for economic reasons, but not before the benefits of industrial frequency ac had been demonstrated. These pictures are included because of the historic importance of this line. *Gavin Morrison*

Below: The Midland services had their own bridge across the River Lune at Lancaster, and the set with No M28222M leading has just left Green Ayre station for Heysham on 9 October 1965. Power-operated sliding doors are nowadays taken for granted, but on these trains and others of similar vintage, the sliding doors were opened by hand. *Gavin Morrison*

Below: The first part of the London Midland Region's main line electrification was between Manchester Piccadilly and Crewe. It opened on 12 September 1960. On this original section, Class 304/1 No 010 is forming the 15.20 Altrincham-Crewe on the rather wet 29 August 1980. The first 15 of these units have compartments instead of an open saloon in the Motor Brake Second, which gives them a slightly higher seating capacity than the subsequent Class 304/2. *Martin Higginson*

Left: The 11.32 from Coventry on 22 March 1980 has just arrived at Birmingham New Street, formed of unit No 304.025. These vehicles are non-gangwayed, but lavatories are provided in two of the four cars. Passengers thus have to make a conscious decision before joining the train! *Brian Morrison*

Above: There are few stopping services nowadays up the Trent Valley main line. The 13.40 Stafford-Nuneaton gets away from Tamworth on the up slow on 24 January 1976, consisting of Class 304/2 unit No 030. These trains are powered by four BTH 155kW traction motors. *Les Nixon*

Right: The line between Wolverhampton and Birmingham abounds in sharp curves; the heavy Inter-City traffic causes extensive and expensive track renewals as a result. Less hard on the track is unit No 028, with a Birmingham-bound all-stations service, seen here to the north of Smethwick Rolfe Street on 29 June 1977. *John Glover*

Bottom right: The final Class 304 version is designated 304/3 and has minor detail differences. Unit No 042 enters Bescot with a Walsall-Birmingham New Street train on 27 June 1977. All the Class 304 units are fitted with Gresley bogies in an attempt to improve riding qualities. *John Glover*

Above: Electrification south to London necessitated a build of 50 four-car units for outer suburban use and semi-fast services to Birmingham. The Class 310 units were built from 1965, and have been widely acclaimed both by operators and passengers. An unidentified unit was photographed entering Berkhamsted station alongside the Grand Union Canal with a Euston-Bletchley service in April 1976. *John Glover*

Below: Even with the light catenaries of ac electrification, junctions such as Rugby present a veritable maze of structures and wires, although it is fair to point out that a present day installation would make use of span wires to support the catenary rather than girders. Class 310 No 053 leaves the station on a Birmingham local train on 16 June 1979. *Les Nixon*

Above: Unit No 073 enters Adderley Park with a Birmingham-Coventry service on the afternoon of 4 August 1980. Even the relatively modern Class 310 units are becoming ready for refrubishing, and the first is due to enter Wolverton Works in 1982. This work will include regearing the units for a maximum speed of 90mph instead of their present 75mph maximum. *John Glover*

Left: No 310.069 leaves Birmingham New Street on 22 March 1980 with a Coventry service. New Street is one of the best-sited city stations in the country. As can be seen from this picture, it is adjacent to the Bull Ring shopping centre in the background. *John Glover*

Right: A coat of paint, and appearance change! No 312.203 is now in the current corporate livery. A considerable improvement in looks has resulted through painting the window surrounds black, compared with the bare metal look as delivered. A Birmingham-bound train approaches Stechford on 4 August 1980. *John Glover*

Birmingham
312203

Scottish Region

Left: Until the delivery of the Class 314 units in 1979, Scottish Region's suburban electric services were all worked by units of a similar type. Three Class 303 ac units go to make up this picture at Milngavie on 21 August 1980. 'Trans-Clyde' is the brand name given to 'your interlinking integrated transport system'; 'GG' refers to the Greater Glasgow (now Strathclyde) Passenger Transport Executive. *Peter Ashton*

Below: Bowling station buildings are of an unusual wooden construction, and the whole station is impeccably kept. At the platform end is a small modern signalbox; large power boxes were not in vogue at the time electrification was carried out in 1960, and this is a typical design of the time. No 303.065 is arriving on 2 September 1980 with a Balloch-Springburn service. *Martin Higginson*

Left: The single-track Balloch branch enjoys a half-hourly train service, but the final half mile to the pier station is used only in connection with Loch Lomond sailings. Class 303 unit No 303.027 has just arrived at Balloch Pier on 21 August 1980 over what must be the only stretch of ac electrified line to carry but an occasional train service, which is seasonal at that. *Peter Ashton*

Below: Overhead electrification does not automatically require the installation of colour-light signals. Semaphores with miniature arms are retained at Milngavie, albeit surrounded by protective wire cages. No 303.032 is arriving from Airdrie on 2 September 1980, and has its blind already changed for the return journey. Since the opening of the Argyle line, additional blinds have been fitted in many units at the top of the centre cab window so that route information can be shown. *Martin Higginson*

Above: Airdrie itself has been completely rebuilt, and No 303.027 will shortly be forming the 15.22 to Helensburgh on 22 August 1980. To the right is No 303.061. Electric services terminate here although the line continues for freight traffic to Edinburgh over the former North British line via Bathgate. Broadly speaking, the first 66 electric units are allocated to Hyndland depot together with the new Class 314s; the remainder, including all the Class 311s, are based on Shields. *John Glover*

Below: The long platforms at Glasgow Central are a favourite vantage point for photographers, and Nos 303.083 and 303.077 were captured there on 24 July 1980. It was the North Side services which suffered the indignity of a suspension of electric services due to equipment problems a mere 41 days after their inauguration in 1960. Electric services were not resumed for 11 months, and this delayed the start of these South Side services until 1962. *Cyril Lofthus*

Above: Ninety-one of the original Pressed Steel sets were built, but these were insufficient to meet the needs on the Gourock and Wemyss Bay services which were added to the electric network in 1967. An additional 19 similar units of Class 311 were constructed by Cravens. However, all units are interchangeable, and this view is of No 303.072 on a service from Gourock leaving Paisley Gilmour Street for Glasgow Central on 21 August 1980. *John Glover*

Below: A Gourock-Glasgow train is seen near Bogston on 2 September 1980 as the evening mist comes in from the Clyde and begins to obscure the shipyard in the background. One of the pleasures of travel on Glasgow's electric services is the tremendous variations of the views from the train in quite short distances. *Martin Higginson*

Above: The Wemyss Bay branch is single track, and No 303.085 is arriving at the attractive Inverkip station with a train for Glasgow on 21 August 1980. The design of these units was overseen by British Rail's Design Panel; as can be seen, they originally had wrap-round windscreens similar to the LMR Class 310 units, and the Class 309 Clacton stock of the ER. *John Glover*

Below: The pleasant small town of Gourock is also a port, and there are many steamer sailings. During the summer months there is a comprehensive range of rail/sea excursions from Glasgow. No 303.072 has just arrived from Glasgow on 20 January 1980, but there are no holidaymakers at this time of year. *Brian Morrison*

Top left: Both the Class 303 and 311 units share an interior design which has proved popular with passengers. The good views to be had through the front windows are enhanced for most travellers in the end coaches of a set by the low-backed but comfortable seats. This picture is of a Class 311 unit, which differs internally from the Class 303 only in that it is fitted with fluorescent lighting instead of conventional bulbs. *John Glover*

Centre left: The guard looks on patiently as the photographer dashes out to get a picture of a Glasgow-Neilston train at Whitecraigs on 20 August 1980 and then rushes back again. With sliding doors such expeditions are more hazardous than with slam doors which can be left on first catch! As is again evident here, the general standard of upkeep of so many Scottish Region stations is a credit to their staffs. *John Glover*

Below: A Cathcart Circle train departs from Pollockshields East on 22 August 1980 formed of unit No 311.098. Nearly all these trains had been repainted in blue and grey livery by the end of 1980 and all were carrying six-digit numbers. Refurbishing had yet to start, and this was programmed from mid-1981 at Glasgow Works. *Peter Ashton*

Above: East of Partick, the eastbound Argyle line diverges from the Queen Street line with a burrowing junction, and the tracks come together again here at Finnieston. This was the former Stobcross station of the Caledonian Railway, closed in 1959 and reopened in 1979. The Motherwell train, formed of No 303.019 is running on the remaining single track of the former Lanarkshire & Dumbartonshire Railway. Westbound trains use the old freight spur. The date is 1 September 1980. *Martin Higginson*

Right: As a prototype for the new stock for Glasgow suburban services required with the Argyle line opening, the ex-Southern PEP unit No 920.001 was fitted with thyristor control and transferred to Hyndland depot. Now an ac unit, it underwent evaluation trials in 1976, and is seen here passing Dalreoch on 12 April 1976. *Finlay McKechnie*

Top left: No 314.210 briefly catches the sunlight in the otherwise underground Anderston station with the 15.25 Dumbarton Central-Motherwell on 18 August 1980. This unit was used by HM The Queen when she opened the new line on 1 November 1979. The three-mile tunnel section of the Argyle line is laid with concrete slab track, which greatly eased the problems of providing sufficient clearances for 25kV ac electrification. Part of the original Blue Train network had been at 6.25kV, but this has now all been converted to 25kV. *Cyril Lofthus*

Centre left: Glasgow Central Low Level station was abandoned in 1964. Conversion to enable it to fulfil its new role in the Trans-Clyde network involved lowering the platform level, as can be seen by the base of the pillars, which now have tile surrounds for use as seats. There is also only one island platform now, the remains of the other having been screened off. Unit No 314.205 stands at Central and will shortly leave for Milngavie on 21 August 1980. *John Glover*

Below: Unit No 314.205 is seen again, this time at Motherwell on 22 August 1980. The train is the 15.23 to Dumbarton via Bellshill and Singer. Despite the many attractions of this PEP-derived stock, the view forward is no longer available to passengers. It is perhaps a tribute to the Class 303/311 design that the 314 seems less of an advance than, say, the Class 508s compared with Southern stock of 20 years earlier. *Peter Ashton*

Glasgow Underground

Left: An escalator descends from street level to bring the prospective passenger to the Underground trains at Partick. To those used to more grandiose systems, the short 130ft station tunnels, the narrow 4ft gauge of the track, and the small size of the trains, which have to fit into 11ft diameter tunnels, come as something of a surprise. Car No 119 is seen on the Inner Rail on 22 October 1980. *Cyril Lofthus*

Above: All stations used to be simple island platforms, but the busier ones have had an additional platform constructed. Train No 6 will only open its doors to the platform on the right, thus keeping the two flows of passengers separated. This view was taken on 22 August 1980. *John Glover*

Right: A two-car train arrives at Hillhead on the same day. The stations are all designed to take three-car trains, and some have operated since November 1980. Six trains provide a 3½min frequency on each circle at the peaks, reducing to three trains at 8min frequency in the evenings. With the old system which closed in 1977, the only way to change the service frequency was to lift trains bodily into or out of the system. There is now a conventional connection to the surface at Govan. *John Glover*

Right: Buchanan Street is one of the busiest stations, and car No 113 is at the front of this Outer Rail train on 18 August 1980. All the cars are identical, and are painted bright orange with a cream waist stripe. The continuous nature of the service quickly resulted in the line being dubbed the 'Clockwork Orange', though there is nothing of an 'X' certificate nature about the system. *Cyril Lofthus*

Below: Another view of Buchanan Street on 22 August 1980 demonstrates the diminutive size of the trains. A complete circuit of the 15 stations takes 22min to accomplish, and there is a flat fare which was 20p at the time of opening on 16 April 1980. After eight months, patronage had reached 40,000 per day, and with continuing growth will reach the planned target of 15 million per annum. *John Glover*

Above: Two passengers wait at Kelvinbridge for their train on 22 August 1980. This is one of the unrebuilt stations, although it must be said that everything in sight has been renewed. One of the pleasant features is the public address system on which messages for both staff and public are broadcast, resulting in a remarkably homely atmosphere. *John Glover*

Below: Car No 127 was photographed at Govan on 22 August 1980. The high level of the conductor rail can be seen, 12in above the running rails. This is necessary due to the restricted size of the tunnels. The system is operated at 600V dc, and the trains were built by Metro-Cammell in Birmingham. *John Glover*

Tyne and Wear Metro

Left: The various electrification systems possible for the Metro were evaluated, and 1,500V dc overhead was chosen on the basis of detailed cost comparisons. Ironically, in view of the previous electrification which was abandoned in 1963, third rail was ruled out for safety reasons by the Inspecting Officer of Railways. A one-mile test track was built near Shiremoor, and prototype Metrocar No 4001 was photographed there on 17 April 1980. There was an optional section featuring severe reverse curves near the northern end, and this can be seen to the left of the picture. *John Glover*

Bottom left: On the same day, car No 4001 has just passed through the demolished artificial tunnel. It was intended that the test track be used to demonstrate British-built lightweight rolling stock generally, but only the Hong Kong cars have been tested, and they were too big for the tunnel. The substantial fencing is of note. *John Glover*

Right: Inside, the cars are neat and functional. Seats for 84 are provided, and there is space for large items of luggage over the articulated section in the centre. This is prototype car No 4001 which has doors in the ends; these are omitted in the production batches as it was decided to construct continuous walkways in the tunnel sections for detraining passengers in emergency. *John Glover*

Below: Something of a record must have been created by production car No 4077, which was placed in the National Railway Museum at York before it had even entered service. It is seen here on 22 September 1979, when it was being exhibited as part of the Centenary of Electric Railways celebrations. Due to its 90ft length it partially occupied the turntable, which did not endear it to the museum's staff! *John Glover*

Above: The car sheds are situated at South Gosforth in the vee between the Bank Foot and the Coast lines. This depot has been transferred from British Rail. Car No 4004 is leaving for test running between South Gosforth and Haymarket on 18 April 1980. A British Rail dmu can be seen in the background, together with one of Tyne & Wear's Brush-built 427hp diesel shunters. *John Glover*

Left: The entire Metro system is controlled from a nerve centre at South Gosforth station. This includes power supply and station monitoring as well as signalling. However, South Gosforth East signalbox still stands on 18 April 1980 as it is passed by Metrocar No 4029. *Peter Ashton*

Above: A pair of Metrocars pass at West Jesmond on the same day during driver training. This is a former BR station which has been renovated for Metro. When complete, the system will have 41 stations, many of them newly constructed. West Jesmond is on the initial stretch of line opened from Haymarket to Tynemouth on 11 August 1980. *John Glover*

Below: Whitley Bay sees a healthy number of passengers for a midday train to Haymarket on 26 August 1980. Trains travel at a 50mph maximum, and the time taken for the 12-mile throughout journey is 26min with 12 intermediate stops. Single fare when the line opened was 38p, and the service operates at 10min intervals over an 18½-hour day. *Martin Higginson*

Above: Ticketing is by serve-yourself machines, but a PTE official was on hand to advise passengers in difficulty at Tynemouth on 26 August 1980. Besides ordinary Metro tickets, passengers can purchase 'Transfare' single tickets which are available for through journeys by bus and rail. *Martin Higginson*

Below: Although provision is made for disabled persons in wheelchairs, the same cannot be said for illiterates! There are eleven separate notices on the inside of this pair of Metrocar doors, three of which are repeated twice. *Martin Higginson*

Above: Shiremoor is one of the new stations, servicing a modern residential estate. A Tynemouth train is leaving, with unit No 4063 at the front, on 26 August 1980. Operating problems have chiefly concerned the sticking of the plug type doors, which seems to be a problem encountered by many operators. By early 1981 the system was carrying 150,000 passengers a week, and the numbers were rising. *Martin Higginson*

Below: The excellent forward view is being taken advantage of in unit No 4025 as it enters the tunnel section at Jesmond on 26 August 1980. In the background is the single-track link to Manors, the former BR route. The articulated Metrocars can surmount a 1 in 25 gradient, but the steepest on the system as at present planned is 1 in 30 on each side of the new bridge spanning the River Tyne. *Martin Higginson*

Eastern Region

Above: The first line to receive production models of the PEP stock was the Great Northern with its 64 Class 313 inner suburban units. They are 25kV ac/750V dc dual-voltage units; each end vehicle contains four 110hp traction motors, and the centre vehicle carries the pantograph. Unit No 313.059 leads a six-car train into Wood Green on 9 July 1979 with a Moorgate-Welwyn Garden City service. *John Glover*

Below: The tunnels around Hadley Wood have always attracted photographers. No 313.059 heads the 14.56 Welwyn Garden City-Moorgate on 26 July 1979. Since their introduction in 1976, these units have had a poor reliability record with only 7,600 miles between failures. The most important causes were problems with the sliding doors, and the tripcock mechanism used on the underground section. *Brian Morrison*

Top: Also just north of Hadley Wood, No 313.046 is en route for Welwyn Garden City on 19 September 1980. Although it was intended to operate three cars only off-peak, difficulties with the American Tightlok coupler, which can only be operated satisfactorily on straight track, has led to six-car trains running throughout the day. *Gavin Morrison*

Above: In terms of attracting patronage, the Class 313s have been most successful, and about 32% more commuting journeys are being made than there were before electrification. Unti No 313.041 is, however, on a rather different type of service. It is leaving Letchworth on the 17.27 to Moorgate on 16 August 1980. This train runs via Langley Junction and Hertford North, and provides the hourly service over this section. *Les Bertram*

Above: Unit No 313.006 stands at Hertford North waiting to depart with the 12.00 service to Moorgate on 16 November 1980. All stations on these lines have been renovated to a greater or less extent, and Hertford North retains a pleasant Great Northern atmosphere. *John Baker*

Below: In the normal course of events, Class 313 units are seldom seen in passenger service at Kings Cross. They do however appear as parcels trains. This is platform 1 at about 21.00 on 10 March 1980. Fortunately, the large floor area makes it unnecessary for mailbags to be placed on seats, a practice which is still distressingly common elsewhere. *Martin Higginson*

Above: Works attention for the York-built Class 313s is the responsibility of Horwich. No 313.006 is hauled past Heaton Lodge Junction, far from any electrified tracks, by No 31.303 on 24 July 1980. Such trains need to travel as Class 8 freights, as no continuous brakes are available. The prominent roof ventilators of the emu are peculiar to the first few of the class. *Gavin Morrison*

Below: Night time at Kings Cross sees No 312.711 waiting to depart for Royston on 10 June 1980. These 25kV-only machines were built from 1975. Twenty-six are allocated to the Great Northern, and it is acknowledged that this is sufficient to work electric services to Cambridge, if and when electrification is authorised. *Gavin Morrison*

Left: Most Royston trains leave from the former suburban station at Kings Cross, now platforms 9 and 10. No 312.703 is vacating No 10 on the morning of 12 July 1979. Kings Cross locomotive facilities are now disused, and the remainder of the suburban station has had all the track removed. *John Glover*

Above: No 312.026, the last of the batch, hurries through Wood Green with a down train on 14 September 1978. As electric units are often referred to by their last three numbers, all the Class 312/0s had 700 added to their numbers to avoid confusion with other classes. No 312.026 thus became No 312.726. *John Glover*

Centre right: In the open countryside near Ashwell & Morden, No 312.722 forms the 09.50 from Kings Cross to Royston on 5 October 1980. Electrification has stimulated outer suburban traffic too, and peak hour travel towards London from stations beyond Hitchin has risen by about 50% as a result. *John Baker*

Bottom right: Unit No 312.725 has been fitted with a high intensity headlight, which is rapidly becoming standard British Rail practice. It is seen here on the 10.50 Kings Cross-Royston at the automatic half barrier Litlington Crossing on 26 October 1980. *John Baker*

1F
54
623

CHELMSFORD
09

Far left, top: Electrification on the Great Eastern lines dates back to 1949, and four generations of rolling stock were on view at Liverpool Street on 10 July 1980. From left to right are a Class 306 of 1949, a Class 312/1 of 1975, Class 305/2 No 502 of 1960 and Class 309/3 No 623 of 1962. *Martin Higginson*

Bottom left: The Class 306 units were built as 1,500V dc vehicles, and their similarity to the Class 506 Manchester-Hadfield units is obvious. Before conversion to ac, the dc electrification was extended to Southend Victoria and Chelmsford. No 09 – an additional nought was added subsequently to distinguish converted units – stands at Chelmsford on opening day, 11 June 1956. *British Rail*

Left: Besides banishing the pantograph and the guard to the centre vehicle, which also contains the rectifying equipment, conversion entailed replacing the marker lights with a two-digit reporting number. Unit No 011 is forming a local service to Liverpool Street as it calls at Ilford on 10 November 1977. *Peter Ashton*

Below: Class 306 unit No 040 arrives at Liverpool Street to form a down evening peak service on 28 September 1979. At this stage the traction current was still being fed at 6.25kV; the second change in 20 years, this time to 25kV, took place in October 1980. Catenary supports are the originals, installed for 1,500V dc in 1949. *John Glover*

Above: The original unit, No 306.001, leaves Stratford for Shenfield as part of a nine-car formation on 21 July 1980. These units will retain all-blue livery until they are retired on completion of Class 315 deliveries. The built-up roof on the leading car where the pantograph once stood is visible, and the former guard's compartment is now part of the passenger saloon. *John Glover*

Left: Internally, a three-car set offers 168 seats, in spite of there being two pairs of sliding doors in the side of each car. The 91 trains of this class will be replaced by 61 four-car units of Class 315; because of the higher seating capacity of the latter trains, peak frequencies will be reduced when they are fully operational. *Martin Higginson*

Above right: The Class 307 units were built by Eastleigh works for the 1,500V dc overhead system in 1956 and were converted to ac in 1960. No 307.102 runs into Wickford with an afternoon Liverpool Street-Southend Victoria service on 20 May 1980. These units also lost marker lights, but acquired a four-character headcode on conversion to ac. *John Glover*

Right: A few minutes later, No 104 arrives at Wickford in the up direction. The scene has changed little over the years; searchlight signals, manual signalbox, electrical equipment and train are all at least 25 years old. *John Glover*

2D
76
307 102

Wickford
LIVERPOOL St
2D
31
104

Above: The Class 307 units are to be the first Great Eastern emus to be refurbished, although No 103 has merely been repainted. It is arriving at Gidea Park with the 12.25 Southend Victoria-Liverpool Street on 15 January 1981. Electrical equipment is mounted in the second car, an unusual feature for dc traction at the time, but standard ac practice. *Peter Ashton*

Left: The Witham-Braintree branch was a later addition to the GE electrified network, and was energised in 1977. Class 308/1 No 134 is about to leave Witham in September 1978. The 33 units of this class were intended for outer suburban work, and have the raked back front, a Design Panel inspired improvement. *Kevin Lane*

Right: No 308.146 is in home territory on the Colchester-Clacton branch, and is entering Wivenhoe with the 17.56 all stations Colchester-Clacton on 24 June 1980. This train follows the 17.00 from Liverpool Street, providing a connection for intermediate stations to Clacton. *Martin Higginson*

2F06
Clacton
308146
Way o

Above: The Class 302 units monopolise the Fenchurch Street services. They were the first ac electric units, appearing in December 1958, and trial running commenced on the Colchester-Clacton/Walton branch, the electrification of which had been advanced to provide test facilities. Unit No 296 enters Stepney East with a down empty stock train on 20 May 1980. *John Glover*

Centre right: A newly painted No 281 is followed by a scruffy No 285 on the 09.37 Pitsea-Fenchurch Street via Tilbury on 19 November 1979. The train is passing Ripple Lane yard just outside Barking. No destination blinds or headcodes are carried nowadays on Fenchurch Street services. *Brian Morrison*

Bottom right: On the frosty morning of 9 February 1978, Class 302 No 301 enters Dagenham Dock station with a Fenchurch Street train. The Class 08 diesel shunter in the background shunts traffic for the adjacent Ford plant. The Class 302 units are all mounted on Gresley bogies, which give an acceptable ride if maintained properly. *John Glover*

Left: Basildon station was opened in 1974 after a long fight by the local authority, which felt that both Laindon (1½ miles) and Pitsea (2¼ miles) were too distant from the town centre. The municipal style of the station buildings is in marked contrast to British Rail's standard productions. Unit No 261 is at the head of a Shoeburyness train on 20 May 1980. *John Glover*

Below: The 15.30 Shoeburyness-Fenchurch Street approaches Leigh-on-Sea on 28 September 1980, composed of units Nos 283 and 262. Views of the sea from the railway are limited, but the boat in the picture helps to convey the right atmosphere. *Les Bertram*

244
9Z
314

Far left, top: Twenty-six trains for the Manchester-Bury service was an absurd over-provision, so when a mishap caused the loss of a Driving Trailer from Class 302 unit No 244, No M77164, a stored Bury line trailer, was converted to take its place. It is seen here at the rear of a train leaving Upminster on the down line on 20 May 1980. *John Glover*

Far left, bottom: In 1961 a further nine train sets were built for the LT&S services. When new, these all incorporated a luggage van for parcels traffic. Subsequent reappraisal of requirements led to the conversion of vans on four of the units to passenger accommodation in 1971. Class 308/4 No 314, one of the converted units, leaves Upminster with a down train on 20 May 1980. *John Glover*

Top left: One of the remaining luggage vans – strictly a Motor Luggage Van as it carries the traction equipment – is marshalled into Class 308/2 unit No 320. It is seen at Pitsea on the same day. Eastern Region practice contrasts with the Southern, who have their separate Class 419 Motor Luggage Vans. *John Glover*

Centre left: This is the interior of the converted van on unit No 314. The seating is reminiscent of that used on the Class 310s, perhaps not surprisingly as the work was carried out at Wolverton. Window arrangement is however in Mark 1 locomotive-hauled stock style. *John Glover*

Below: Local services in North East London are operated by the 52 three-car units of Class 305/1 built at York in 1960. Unit No 415 arrives at Broxbourne with a Hertford East-Liverpool Street service via the Lea Valley on 3 October 1980. *John Glover*

2M00
406
DERV
DERV
125P
30P

Left: On 29 November 1980, Unit No 406 brings up the rear of a Bishops Stortford-Liverpool Street train as it descends the bank from Bethnal Green towards the terminus. Three similar units, classified 308/3, were subsequently added to the fleet. These have English Electric instead of GEC equipment. *Les Nixon*

Right: Internally, these trains consist of open saloons, and there are plenty of windows. The intermediate partitions are being removed; they were intended to segregate smoking and non-smoking accommodation, but the increase in non-smoking space in recent years has made them redundant. *GEC*

Below: 'With a pair of opera glasses, you could see to Hackney Marshes . . .' and a Class 305/1 unit on the 10.40 Liverpool Street-Chingford. It is crossing the original Cambridge main line and a River Lea flood channel on 3 October 1977. This location is better known as Coppermill Junction. *Martin Higginson*

Above: The 19 Class 305/2 units were built at Doncaster in 1959. They are almost identical to Class 308 apart from having GEC electrical equipment. Unit No 508 leaves the limit of electrificiation at Bishops Stortford on 2 August 1980 with the 17.04 departure for London. *Les Bertram*

Left: The Class 305/2s, like other GE/LTS units, may be seen anywhere on those systems, but are usually to be found in North East London. Unit No 502, resplendent in newly applied blue and grey, arrives at Sawbridgeworth with an up train on 3 October 1980. As outer suburban units, these trains include first class seating. *John Glover*

Above right: Sawbridgeworth has a typically East Anglian feel to it, although it is on the Hertfordshire/Essex border. No 501 arrives with a down train during the afternoon of 3 October 1980. Although the four-character headcodes are mostly disused, the destination blind on the Great Eastern is usually a reliable guide. *John Glover*

Right: The same day sees an unidentified Class 305/2 unit alongside the Stort Navigation between Harlow Mill and Sawbridgeworth. Urbanisation has yet to envelop this area, but the river sees many pleasure craft in summer. *John Glover*

SAWBRIDGEWORTH
501
Passengers must not cross the line

Right: Until the advent of the APT-P, the Clacton units were the only true express ac emus on British Rail, and were intended to be the forerunners of a fleet. Fifteen four-car and eight two-car units were delivered by York in 1962. Unit No 626 is one of the four-car sets without a buffet car, classified 309/3. It was photographed at Colchester on 8 April 1980, where it had stopped on top of the aws ramp. *Peter Ashton*

Below: Unit No 626 is seen again, this time at the head of a 10-car Liverpool Street-Clacton train passing Bethnal Green on 29 November 1980. All train sets are fitted with four GEC 222hp traction motors; the higher power/weight ratio in the two car sets was intended to enable fast timings to be maintained in rush hours, when they are attached to a pair of standard four-car units. *Les Nixon*

Above: Class 309/3 No 623 arrives at Shenfield on 1 May 1978 with the 12.50 from Liverpool Street. This train will proceed to Thorpe-le-Soken where the train will divide. The front four vehicles are for Walton-on-Naze and the rear four, including one of the celebrated Griddle cars, are for Clacton. All buffet facilities are now withdrawn from the route, but the destination boards were still in place on the coach sides and in use. *Les Bertram*

Centre right: The two-car units have no first class accommodation, and are rarely seen on their own. Class 309/1 No 602 approaches Gidea Park at the head of an up afternoon service on 20 May 1980. All these 100mph trains are fitted with heavy Commonwealth bogies to improve riding characteristics, but at the expense of a weight for a two-car unit of 99 tons. The Class 309 as a whole are the most reliable units on the Eastern Region. *John Glover*

Bottom right: Two-car unit No 603 is at the head of the 12.26 Clacton-Liverpool Street as it passes the long closed station of Thorrington on the Clacton branch on 29 June 1980. All Class 309 units have now had their wrap-round windscreens replaced, which has not improved their appearance. *John Baker*

Below: To increase capacity, half of the two-car units had two intermediate vehicles inserted between the Driving Motor Brakes and the Driving Trailers in 1974. They are now Class 309/4, and unit No 608 is at the front of the 11.50 Clacton/11.47 Walton-on-Naze-Liverpool Street, seen near Marks Tey on 4 October 1980. The additional vehicles are a Mark I locomotive-hauled composite (CK) and a compartment second (SK). *Les Bertram*

Bottom: Nineteen Class 312/1 units are allocated to the Great Eastern lines, and are used principally on the stopping trains to Colchester. The first of the class, No 312.101, passes Stratford with an up train on 24 October 1980. Official renumbering to 312.781 will be carried out eventually, but in the meantime paper stickers in the centre cab window show the revised number. *Cyril Lofthus*

Right: The Class 312 units have a neat driver's desk arrangement, with virtually everything within reach from a sitting position. Not so long ago, a seat was considered an optional extra, and one is reminded of the early days of motorbus operation when the Metropolitan Police even objected to windscreens! *John Howie*

Above: Several new trains for the Gidea Park services were lined up in the recently electrified carriage sidings at Stratford on 15 January 1981, awaiting commissioning. Nearest the camera are Nos 315.816 and 315.819. Comprehensive destination blinds seem to have been fitted, with both Witham and Leigh included. *Peter Ashton*

Left: Later the same day, No 315.811 passed Gidea Park with a driver training trip from Southend Victoria to Ilford Car Sheds. These are four-car units arranged for 25kV only. Interestingly, they have been fitted with four-character headcode blinds, which see little use nowadays on British Rail. The fleet consists of 61 Class 315 units, bringing the total of production cars based on the PEP prototypes to 755. *Peter Ashton*

London Transport

Right: The celebrated 1938 tube stock of London Transport is now largely confined underground. Working only on the Bakerloo Line, trains come to the surface at Queens Park, where most terminate. Seen from the overbridge at Queens Park station, an Elephant & Castle-bound train is leaving while a northbound train emerges from the tunnel on 22 July 1980. On each side of the Bakerloo Line can be seen the tracks of the LMR dc lines to Euston, whilst on the right are the ac lines and a Class 86 with empty stock. *John Glover*

Below: Before the Jubilee Line opened in 1979, Bakerloo trains also worked to Stanmore. A southbound train leaves Wembley Park in December 1977. The design of the 1938 stock was based on experimental vehicles built in 1935, and 1,121 cars of the production design had been delivered by Metro-Cammell by the outbreak of war. Thirty-four seven-car sets remain; these have received Extra Heavy Overhauls (hence EHO stock) but all were due to be replaced in the mid-1980s. *Kevin Lane.*

Below: The four trains required for the Watford Junction-Elephant & Castle service are stored overnight at BR's Croxley Green depot. Two of the trains were photographed alongside a native Class 501 unit on 29 October 1978. The all-day London Transport service provided north of Queens Park a few years ago is no more, although since the Jubilee Line opened certain trains now operate from the south through to the new Stonebridge Park LT depot. *Brian Morrison*

Bottom: The three 1956 stock prototypes were a post-war development of the 1938 stock. Apart from aluminium bodywork, the new cars incorporated fluorescent lighting, rubber springing, rearranged seating and door fault lights. They were sent to the Piccadilly Line, and this vintage photograph shows a westbound train at Piccadilly Circus. Both the advertisements and the style of dress suggest that the picture was taken when the trains were new. *London Transport*

Left: Following trials with the 1956 stock, enough cars for the complete re-equipment of the Piccadilly Line were ordered from Metro-Cammell of Birmingham. A number of cars of what became known as 1959 stock are here having the finishing touches put to them. The production vehicles are immediately distinguishable from the prototypes as it was decided to discard the five marker lights and replace them with only two. *London Transport*

Below: With the delivery of the 1973 tube stock, the 1959 stock trains were transferred to the Northern Line. A northbound seven-car train is approaching Golders Green on 2 March 1978. The complex protection arrangements on the track to ensure that pick-up shoes are not damaged by point and crossing work are particularly prominent. *John Glover*

Top: One of the few single-track parts of the Underground is the Mill Hill East branch, seen here with a 1959 stock train arriving at the wooden platformed terminus on 14 September 1978. This formed part of the frustrated plans to extend the Northern Line under the 1935-40 New Works Programme. Electrification to Mill Hill East was completed in 1941, but only because of the nearby barracks. Passengers for Edgware have to catch a bus outside. *John Glover*

Above: The 1960 stock built by Cravens was intended as a prototype for Central Line replacement, but a slight variant of the 1959 stock was adopted in the event. Some of the 1960 cars had two silver painted pre-1938 trailers placed between them and were converted for automatic train operation (ato). They are thus operated on the Woodford-Hainault branch of the Central Line. A Hainault service leaves Roding Valley on 23 February 1979. *John Glover*

WEST RUISLIP

Far left, top: A 1962 stock train passes Ruislip depot on 9 July 1980. Spare Central Line trains await their next duties alongside withdrawn District Line CO/CP stock awaiting removal for scrap. The semaphore signal refers to a Marylebone-Banbury dmu which had just passed on the ex-Great Western Birmingham main line; with the removal of the LT semaphores at North Weald and the conversion of Richmond SR-owned station to colour-lights, semaphores are now extinct as far as LT trains are concerned. *Martin Higginson*

Far left, bottom: At North Acton Junction the line to Ealing Broadway diverges to the left. The 1962 stock trains in this picture are proceeding to and from West Ruislip. The approaching train is about to reach the unworked trailing points at the junction, which will be set correctly by the action of the wheel flanges. This feature is not found on British Rail because of the danger that a loose-coupled freight train might accidentally set back and land 'on the floor'. The date is 3 October 1977. *John Glover*

Left: When the Central Line was extended west from White City, the tracks were arranged for right hand running due to the previous existence of a terminal loop. Left hand running is resumed half a mile further on towards East Acton. This view, taken on 6 February 1980, shows a West Ruislip-bound train of 1962 stock crossing the flyover. *John Glover*

Below: The extension of the Central Line under Eastern Avenue comes to the surface at Newbury Park. Formerly, services in this area were provided by the LNER by means of a branch from Ilford. LNER passenger services ceased in 1947, and freight in 1956. The abandoned link to the GE main line is in the centre of this picture, taken on 3 October 1980. *John Glover*

Left: London Transport's most remote outpost is the six mile Epping-Ongar branch. It is operated by a four-car shuttle of 1962 tube stock, seen here approaching Blake Hall en route for Epping on 4 March 1977. This was a part of the LNER legacy which London Transport was perhaps unwilling to accept. Certainly, it was not electrified until 1957, eight years after electric services reached Epping. *John Glover*

Right: Duties on the Woodford-Hainault line are shared by the 1960 stock and the 1967 Victoria Line stock. A train of the latter leaves the long platforms at Chigwell for Hainault on 3 October 1980. Some 1967 stock is always kept at Hainault depot for this service, giving the curious a chance to ride on these trains in daylight. *John Glover*

Below: Two trains of 1967 stock pass at the western end of Chigwell station on 3 October 1980. No destination blinds are carried by these units when on the Central Line, presumably because their stay is only temporary. These were the first tube vehicles to be fitted with headlights; the light at the side of the blind aperture is for use at the rear of a train for 'calling on' the following train when assistance is required in the rear. *John Glover*

Left: Internally, this Victoria Line stock Driving Motor car is similar to all tube stock built from 1935 until the advent of the 1973 cars. Innovations in this particular build were the door window glass extending into the roof curvature and the provision of a small 'standback' by the doors. Quality of ride is not good, and this seems to be a persistent problem in tube stock. *John Glover*

Below: Similar trains, but arranged for two-man operation and of seven cars instead of eight, were delivered to the Northern Line as 1972 Mk 1 tube stock. A train of these cars is seen approaching Burnt Oak on 2 March 1978. Thirty trains to this design were built by Metro-Cammell, who have been responsibile for the majority of London Transport stock. *John Glover*

Above: The High Barnet branch of the Northern Line was also former LNER property. West Finchley was opened in 1933 using equipment from closed stations elsewhere on the LNER. The wooden platforms have not been altered in height since the Underground trains took over, and consequently there is a pronounced step down into the trains. This view of a southbound 1972 Mk 1 stock train was taken in May 1974. *John Glover*

Left: The 1972 Mk II build of 33 trains was intended for the Jubilee Line; a southbound train of this stock is arriving at Green Park on 30 September 1980. The station decor is brighter and more cheerful than that used on the Victoria Line, which appears rather dull in comparison. *John Glover*

Above: At Finchley Road the Jubilee (née Bakerloo) Line comes to the surface in between the Metropolitan Line platforms. A Stanmore-bound train of 1972 Mark II stock is arriving on 30 September 1980. These trains are instantly recognisable by their red doors and LT roundels on the car sides; they were also built with future one-man operation in mind. *John Glover*

Left: The Jubilee Line parallels the Metropolitan until Wembley Park is reached, but Metropolitan Line trains do not call at the intermediate stations. This 1972 Mk II stock train is leaving Neasden on 7 July 1978 when it was still on the Bakerloo. *Ian Cowley*

Right: Boston Manor station is a mixture of original architecture at platform level and London Transport above. A 1973 stock train arrives on a Heathrow Central service on 30 September 1980. These six-car trains replaced the 1959 stock which is now on the Northern; each of the new cars is longer, so total train length remains the same. *John Glover*

HEATHROW

HEATHROW
356

Left: Trains of 1973 stock are seen at the east end of Hammersmith on 5 June 1980, between the District Line tracks. Articulation was considered for this stock and experiments were carried out at Acton Works in 1970. However, articulation results in an even greater body overthrow on curves, to the extent that it would probably have been necessary to resort to seven cars again. There are also operational disadvantages. *John Glover*

Right: At Southgate trains from the Cockfosters direction approach over a viaduct and immediately enter a tunnel in which the station is situated. This is the only bored tube tunnel station from which daylight is visible from the platforms. A 1973 stock train nears the tunnel entrance on 5 March 1978. *John Glover*

Below: The CO/CP surface stock of the District Line is contemporary with the 1938 tube stock. Both types are the last London Transport trains to be finished in the familiar red livery, though the shade changed from 'train red' to the brighter 'bus red' in later years. A train of this stock for Tower Hill arrives at the busy platform 2 at Earls Court on 4 June 1980. *John Glover*

Above: A Hammersmith-Whitechapel train of CO/CP stock enters Westbourne Park station on 21 March 1970. Six positions were available for headcode displays, which gave scope for 65 variations. All trains now display two lights in the bottom positions, as the use of headcodes was abandoned from January 1978. *Wyn Hobson*

Left: Use of oil tail lamps still continues with CO/CP and R stock trains as with British Rail's 4SUB units. This seeming anachronism arises because none of these units carries an emergency battery, and it is imperative that a back-up source of lighting is available in case of failure. A train of CO/CP stock stands at Richmond on 28 February 1980. *Colin Marsden*

Above: Diesel shunter No 08.335 takes some condemned CO/CP stock on its last journey from Rotherham Masborough into C. F. Booth's scrapyard on 29 July 1980. As London Transport rolling stock has no side buffers, a match wagon must always be used if it is to be moved by BR locomotives. *Les Nixon*

Left: A train of R stock cars approaches West Kensington forming a District Line train for Ealing Broadway on 4 August 1978. Cars for these trains were built over a 20-year period, the last being constructed in 1959. They are a mixture of aluminium-bodied and all-steel stock, the latter being painted to match. The leading vehicle is one of the original cars built by the Gloucester company. *John Glover*

Above: Typical of the sub-surface stations in Central London is Blackfriars, seen with an R stock short working to Tower Hill on 14 January 1980. The R stock trains will be withdrawn on completion of deliveries of the D stock. *Martin Higginson*

Right: An R stock Driving Motor interior shows the neat method of incorporating the strip lighting, and the guard's controls in the end of the car. The full size proportions of these vehicles compared with tube stock is obvious. *John Glover*

Far right, top: For the electrification to Amersham, some outer suburban trains were required. The A60 and A62 stock vehicles are arranged with transverse seating throughout on a 3+2 arrangement. These aluminium-bodied units monopolise the Amersham and Uxbridge line services. This busy scene is at Baker Street in the evening rush hour of 26 July 1980. *Martin Higginson*

Far right, bottom: An early afternoon Baker Street-Uxbridge train of A60 stock arrives at the typically well preserved and neat gardened station of Ruislip on 9 July 1980. The roof on the footbridge, although no doubt welcomed by passengers, appears to be an architectural afterthought. *Martin Higginson*

WATFORD
BAKER STREET
5044

UXBRIDGE
RUISLIP

Above: The Chalfont & Latimer-Chesham shuttle arrives at the Buckinghamshire terminus in June 1980. There is no need for a four-car train to pass the signal protecting the end of the line, which is cleared only for the eight-car trains which work to and from London in the peak hours. *Kevin Lane*

Below: East London Line services have been worked by A60 stock since June 1977, and on 13 May 1980 a New Cross-bound train has just passed Canal Junction where the New Cross Gate line diverges. Although the line is double track throughout, all three termini consist of a single platform only. *John Glover*

Above: Today's Hammersmith & City services are provided by trains of C69 and C77 stocks, which are almost identical. West of Royal Oak, the Metropolitan Line tracks dive under the Western Region main line and emerge just short of Westbourne Park. A C stock train ascends the ramp out of the short tunnel as a Class 47-hauled express makes for Paddington on 18 September 1979. *John Glover*

Below: C stock crosses Wood Lane by the BBC Television Centre with a Hammersmith-bound train on 6 February 1980. This design is one of the most reliable on London Transport. An unusual feature is that the Driving Motor cars and Trailers are of different lengths, the Driving Motors being 3ft 7in longer to accommodate the driver's cab and thus leaving the passenger saloons the same in each case. *John Glover*

Right: A C69 stock train arrives at Sloane Square on an Inner Rail Circle Line service. This was the first stock to have a sliding door provided for the driver, and it can be seen propped half open on the hot day of 5 June 1980. The passenger saloons have four pairs of double doors on each side, and as a result there are only 32 seats in each 49ft saloon. The interior of this stock is somewhat stark. *John Glover*

Below: At the west end of Whitechapel station, a C stock train leaves for Hammersmith and a D stock train arrives for Upminster on 30 September 1980. The D stock cars are the longest yet seen on London Transport, and are about 8ft longer than a traditional car. The first train was delivered from Metro-Cammell in May 1979, and entered passenger service in early 1980. *John Glover*

Above: A D stock train forming a Richmond service arrives at West Kensington on 5 June 1980. These cars weigh no more than the 1973 tube stock as they have aluminium frames; tube cars need to retain steel underframes as there is not enough space to use aluminium of sufficient section to prevent the car body from flexing – and the doors from jamming. *John Glover*

Left: This interior view of a D stock car shows the passenger controls for opening the four single-leaf 3ft 6in doors each side of the car. There are 44 or 48 seats in each saloon, in a mixture of transverse and longitudinal seating. The flooring remains the traditional maple wood type, which has defied all attempts to replace it with floor tiles or carpeting. *John Glover*

Above: Although electric vehicles are noted for their long lives, few can claim an ancestry back to 1903. This is Electric Sleet Locomotive No ESL 100, which is formed of two old Central London Railway Driving Motor cars joined back to back, with additional unpowered bogies to carry de-icing equipment. The conversion was carried out at Acton Works in 1939/40. It was photographed at Neasden in December 1977. *Kevin Lane*

Below: A more modern sleet unit was built out of two Metropolitan Line T stock Driving Motor cars in 1961. As each only has a cab at one end, these vehicles always run as a pair and are numbered ESL 118A and ESL 118B. The scene is Neasden, December 1977. With the fitting of passenger cars with de-icing equipment, the use of sleet locomotives is diminishing. *Kevin Lane*

The Best of Modern Practice

Top: In July 1980 an exhibition organised by the Fellowship of Engineering took place at Ruislip LT depot and selected examples of modern rolling stock were exhibited. British Rail provided No 508.029 and six cars of APT-P No 370.007 to represent electric traction. *Martin Higginson*

Above: No 508.029 is seen from the other end, together with Tyne & Wear Metrocar No 4055, which was brought down from Newcastle by British Rail. On the right is London Transport's Tunnel Cleaning Train. This unit is formed out of two converted 1938 tube stock Driving Motors and three specially built intermediate vehicles. *Martin Higginson*

Above: A remarkable line-up of front ends with (from left to right) LT's Tunnel Cleaning Train, Tyne & Wear car No 4055, Southern Region's No 508.029 and APT-P No 370.007. *Martin Higginson*

Below: A specially prepared LT 1973 stock train formed the backdrop for the Fellowship of Engineering dinner held in the lifting shop at Ruislip on 9 July 1980. One is reminded of the banquets held to mark the opening of new railways in the last century. Did the dinner coincide with a new 'Age of the Train'? *Martin Higginson*